This planner belongs to :

_____

2026

January
S	M	T	W	T	F	S
				1	2	3
4	5	6	7	8	9	10
11	12	13	14	15	16	17
18	19	20	21	22	23	24
25	26	27	28	29	30	31

February
S	M	T	W	T	F	S
1	2	3	4	5	6	7
8	9	10	11	12	13	14
15	16	17	18	19	20	21
22	23	24	25	26	27	28

March
S	M	T	W	T	F	S
1	2	3	4	5	6	7
8	9	10	11	12	13	14
15	16	17	18	19	20	21
22	23	24	25	26	27	28
29	30	31				

April
S	M	T	W	T	F	S
			1	2	3	4
5	6	7	8	9	10	11
12	13	14	15	16	17	18
19	20	21	22	23	24	25
26	27	28	29	30		

May
S	M	T	W	T	F	S
					1	2
3	4	5	6	7	8	9
10	11	12	13	14	15	16
17	18	19	20	21	22	23
24	25	26	27	28	29	30
31						

June
S	M	T	W	T	F	S
	1	2	3	4	5	6
7	8	9	10	11	12	13
14	15	16	17	18	19	20
21	22	23	24	25	26	27
28	29	30				

July
S	M	T	W	T	F	S
			1	2	3	4
5	6	7	8	9	10	11
12	13	14	15	16	17	18
19	20	21	22	23	24	25
26	27	28	29	30	31	

August
S	M	T	W	T	F	S
						1
2	3	4	5	6	7	8
9	10	11	12	13	14	15
16	17	18	19	20	21	22
23	24	25	26	27	28	29
30	31					

September
S	M	T	W	T	F	S
		1	2	3	4	5
6	7	8	9	10	11	12
13	14	15	16	17	18	19
20	21	22	23	24	25	26
27	28	29	30			

October
S	M	T	W	T	F	S
				1	2	3
4	5	6	7	8	9	10
11	12	13	14	15	16	17
18	19	20	21	22	23	24
25	26	27	28	29	30	31

November
S	M	T	W	T	F	S
1	2	3	4	5	6	7
8	9	10	11	12	13	14
15	16	17	18	19	20	21
22	23	24	25	26	27	28
29	30					

December
S	M	T	W	T	F	S
		1	2	3	4	5
6	7	8	9	10	11	12
13	14	15	16	17	18	19
20	21	22	23	24	25	26
27	28	29	30	31		

Year in Pixels

	J	F	M	A	M	J	J	A	S	O	N	D
1.												
2.												
3.												
4.												
5.												
6.												
7.												
8.												
9.												
10.												
11.												
12.												
13.												
14.												
15.												
16.												
17.												
18.												
19.												
20.												
21.												
22.												
23.												
24.												
25.												
26.												
27.												
28.												
29.												
30.												
31.												

Color Codes

Notes

January 2026

MONDAY	TUESDAY	WEDNESDAY	THURSDAY
			1
5	6	7	8
12	13	14	15
19	20	21	22
26	27	28	29

January

FRIDAY	SATURDAY	SUNDAY	NOTES
2	3	4	○
			○
			○
			○
			○
9	10	11	○
			○
			○
			○
16	17	18	○
			○
			○
			○
23	24	25	○
			○
			○
			○
			○
30	31		NOTES

February

MONDAY	TUESDAY	WEDNESDAY	THURSDAY
2	3	4	5
9	10	11	12
16	17	18	19
23	24	25	26

February 2026

FRIDAY	SATURDAY	SUNDAY	NOTES
		1	○
			○
			○
			○
			○
6	7	8	○
			○
			○
			○
13	14	15	○
			○
			○
			○
			○
20	21	22	○
			○
			○
			○
			○
27	28		NOTES

March 2026

MONDAY	TUESDAY	WEDNESDAY	THURSDAY
2	3	4	5
9	10	11	12
16	17	18	19
23	24	25	26

March 2026

FRIDAY	SATURDAY	SUNDAY	NOTES
		1	○
			○
			○
			○
			○
6	7	8	○
			○
			○
			○
13	14	15	○
			○
			○
			○
			○
20	21	22	○
			○
			○
			○
			○
27	28	29	30 / 31

April

MONDAY	TUESDAY	WEDNESDAY	THURSDAY
		1	2
6	7	8	9
13	14	15	16
20	21	22	23
27	28	29	30

April

2026

FRIDAY	SATURDAY	SUNDAY	NOTES
3	4	5	○
			○
			○
			○
			○
10	11	12	○
			○
			○
			○
			○
17	18	19	○
			○
			○
			○
24	25	26	○
			○
			○
			○
			○
			NOTES

May

MONDAY	TUESDAY	WEDNESDAY	THURSDAY
4	5	6	7
11	12	13	14
18	19	20	21
25	26	27	28

May 2026

FRIDAY	SATURDAY	SUNDAY	NOTES
1	2	3	○
			○
			○
			○
			○
8	9	10	○
			○
			○
			○
15	16	17	○
			○
			○
			○
			○
22	23	24	○
			○
			○
			○
			○
29	30	31	Notes

June

MONDAY	TUESDAY	WEDNESDAY	THURSDAY
1	2	3	4
8	9	10	11
15	16	17	18
22	23	24	25
29	30		

June 2026

FRIDAY	SATURDAY	SUNDAY	NOTES
5	6	7	○
			○
			○
			○
			○
12	13	14	○
			○
			○
			○
19	20	21	○
			○
			○
			○
			○
26	27	28	○
			○
			○
			○
			○
			Notes

July

MONDAY	TUESDAY	WEDNESDAY	THURSDAY
		1	2
6	7	8	9
13	14	15	16
20	21	22	23
27	28	29	30

July

2026

FRIDAY	SATURDAY	SUNDAY	NOTES
3	4	5	○
			○
			○
			○
			○
10	11	12	○
			○
			○
			○
			○
17	18	19	○
			○
			○
			○
			○
24	25	26	○
			○
			○
			○
			○
31			NOTES

August 2026

MONDAY	TUESDAY	WEDNESDAY	THURSDAY
3	4	5	6
10	11	12	13
17	18	19	20
24	25	26	27

August

FRIDAY	SATURDAY	SUNDAY	NOTES
	1	2	○
			○
			○
			○
			○
7	8	9	○
			○
			○
			○
14	15	16	○
			○
			○
			○
21	22	23	○
			○
			○
			○
			○
28	29	30	31

September 2026

MONDAY	TUESDAY	WEDNESDAY	THURSDAY
	1	2	3
7	8	9	10
14	15	16	17
21	22	23	24
28	29	30	

September

FRIDAY	SATURDAY	SUNDAY	NOTES
4	5	6	○
			○
			○
			○
			○
11	12	13	○
			○
			○
			○
18	19	20	○
			○
			○
			○
			○
25	26	27	○
			○
			○
			○
			○
			NOTES

October

MONDAY	TUESDAY	WEDNESDAY	THURSDAY
			1
5	6	7	8
12	13	14	15
19	20	21	22
26	27	28	29

October 2026

FRIDAY	SATURDAY	SUNDAY	NOTES
2	3	4	○
			○
			○
			○
9	10	11	○
			○
			○
			○
16	17	18	○
			○
			○
			○
			○
23	24	25	○
			○
			○
			○
			○
30	31		NOTES

November

MONDAY	TUESDAY	WEDNESDAY	THURSDAY
2	3	4	5
9	10	11	12
16	17	18	19
23	24	25	26

November 2026

FRIDAY	SATURDAY	SUNDAY	NOTES
		1	○
			○
			○
			○
			○
6	7	8	○
			○
			○
			○
			○
13	14	15	○
			○
			○
			○
20	21	22	○
			○
			○
			○
			○
27	28	29	30

December
2026

MONDAY	TUESDAY	WEDNESDAY	THURSDAY
	1	2	3
7	8	9	10
14	15	16	17
21	22	23	24
28	29	30	31

December

FRIDAY	SATURDAY	SUNDAY	NOTES
4	5	6	○
			○
			○
			○
			○
11	12	13	○
			○
			○
			○
			○
18	19	20	○
			○
			○
			○
			○
25	26	27	○
			○
			○
			○
			○
			NOTES

December
2025

01 MONDAY

○ _____
○ _____
○ _____
○ _____
○ _____
○ _____
○ _____
○ _____

02 TUESDAY

○ _____
○ _____
○ _____
○ _____
○ _____
○ _____
○ _____
○ _____

03 WEDNESDAY

○ _____
○ _____
○ _____
○ _____
○ _____
○ _____
○ _____
○ _____

04 FRIDAY

○ _____
○ _____
○ _____
○ _____
○ _____

05 FRIDAY

○ _____
○ _____
○ _____
○ _____
○ _____
○ _____
○ _____
○ _____

06 SATURDAY

○ _____
○ _____
○ _____
○ _____
○ _____
○ _____
○ _____
○ _____

07 SUNDAY

○ _____
○ _____
○ _____
○ _____
○ _____
○ _____
○ _____
○ _____

08 MONDAY

○ _____
○ _____
○ _____
○ _____
○ _____

December
2025

09 TUESDAY

10 WEDNESDAY

11 THURSDAY

12 FRIDAY

13 SATURDAY

○ _____
○ _____
○ _____
○ _____
○ _____
○ _____
○ _____
○ _____

14 SUNDAY

○ _____
○ _____
○ _____
○ _____
○ _____
○ _____
○ _____
○ _____

15 MONDAY

○ _____
○ _____
○ _____
○ _____
○ _____
○ _____
○ _____
○ _____

16 TUESDAY

○ _____
○ _____
○ _____
○ _____
○ _____

17 WEDNESDAY

18 THURSDAY

19 FRIDAY

20 SATURDAY

21 SUNDAY

○ _____
○ _____
○ _____
○ _____
○ _____
○ _____
○ _____
○ _____

22 MONDAY

○ _____
○ _____
○ _____
○ _____
○ _____
○ _____
○ _____
○ _____

23 TUESDAY

○ _____
○ _____
○ _____
○ _____
○ _____
○ _____
○ _____
○ _____

24 WEDNESDAY

○ _____
○ _____
○ _____
○ _____
○ _____

25 THURSDAY

26 FRIDAY

27 SATURDAY

28 SUNDAY

December
2025

29 MONDAY

- ○
- ○
- ○
- ○
- ○
- ○
- ○
- ○

30 TUESDAY

- ○
- ○
- ○
- ○
- ○
- ○
- ○
- ○

31 WEDNESDAY

- ○
- ○
- ○
- ○
- ○
- ○
- ○

NOTES

January
2026

01 THURSDAY

○ _____
○ _____
○ _____
○ _____
○ _____
○ _____
○ _____
○ _____

02 FRIDAY

○ _____
○ _____
○ _____
○ _____
○ _____
○ _____
○ _____
○ _____

03 SATURDAY

○ _____
○ _____
○ _____
○ _____
○ _____
○ _____
○ _____
○ _____

04 SUNDAY

○ _____
○ _____
○ _____
○ _____
○ _____

05 MONDAY

○
○
○
○
○
○
○
○

06 TUESDAY

○
○
○
○
○
○
○
○

07 WEDNESDAY

○
○
○
○
○
○
○
○

08 THURSDAY

○
○
○
○
○

09 FRIDAY

○
○
○
○
○
○
○
○

10 SATURDAY

○
○
○
○
○
○
○
○

11 SUNDAY

○
○
○
○
○
○
○
○

12 MONDAY

○
○
○
○
○

January
2026

13 TUESDAY
○
○
○
○
○
○
○
○

14 WEDNESDAY
○
○
○
○
○
○
○
○

15 THURSDAY
○
○
○
○
○
○
○
○

16 FRIDAY
○
○
○
○
○

17 SATURDAY

18 SUNDAY

19 MONDAY

20 TUESDAY

21 WEDNESDAY

22 THURSDAY

23 FRIDAY

24 SATURDAY

25 SUNDAY
○
○
○
○
○
○
○
○

26 MONDAY
○
○
○
○
○
○
○
○

27 TUESDAY
○
○
○
○
○
○
○
○

28 WEDNESDAY
○
○
○
○
○

29 THURSDAY

○ _____
○ _____
○ _____
○ _____
○ _____
○ _____
○ _____
○ _____

30 FRIDAY

○ _____
○ _____
○ _____
○ _____
○ _____
○ _____
○ _____
○ _____

31 SATURDAY

○ _____
○ _____
○ _____
○ _____
○ _____
○ _____
○ _____
○ _____

NOTES

_____
_____
_____
_____

February
2026

01 SUNDAY
- ○ _____
- ○ _____
- ○ _____
- ○ _____
- ○ _____
- ○ _____
- ○ _____
- ○ _____

02 MONDAY
- ○ _____
- ○ _____
- ○ _____
- ○ _____
- ○ _____
- ○ _____
- ○ _____
- ○ _____

03 TUESDAY
- ○ _____
- ○ _____
- ○ _____
- ○ _____
- ○ _____
- ○ _____
- ○ _____
- ○ _____

04 WEDNESDAY
- ○ _____
- ○ _____
- ○ _____
- ○ _____
- ○ _____

February
2026

05 THURSDAY

○ _____
○ _____
○ _____
○ _____
○ _____
○ _____
○ _____
○ _____

06 FRIDAY

○ _____
○ _____
○ _____
○ _____
○ _____
○ _____
○ _____
○ _____

07 SATURDAY

○ _____
○ _____
○ _____
○ _____
○ _____
○ _____
○ _____
○ _____

08 SUNDAY

○ _____
○ _____
○ _____
○ _____
○ _____

February
2026

09 MONDAY

○
○
○
○
○
○
○
○

10 TUESDAY

○
○
○
○
○
○
○
○

11 WEDNESDAY

○
○
○
○
○
○
○
○

12 THURSDAY

○
○
○
○
○

13 FRIDAY

○
○
○
○
○
○
○
○

14 SATURDAY

○
○
○
○
○
○
○
○

15 SUNDAY

○
○
○
○
○
○
○
○

16 MONDAY

○
○
○
○
○

February
2026

17 TUESDAY
○ _____
○ _____
○ _____
○ _____
○ _____
○ _____
○ _____
○ _____

18 WEDNESDAY
○ _____
○ _____
○ _____
○ _____
○ _____
○ _____
○ _____
○ _____

19 THURSDAY
○ _____
○ _____
○ _____
○ _____
○ _____
○ _____
○ _____
○ _____

20 FRIDAY
○ _____
○ _____
○ _____
○ _____
○ _____

February 2026

21 SATURDAY

22 SUNDAY

23 MONDAY

24 TUESDAY

February
2026

25 WEDNESDAY

○
○
○
○
○
○
○
○

26 THURSDAY

○
○
○
○
○
○
○
○

27 FRIDAY

○
○
○
○
○
○
○
○

28 SATURDAY

○
○
○
○
○

March
2026

01 SUNDAY

○ _____
○ _____
○ _____
○ _____
○ _____
○ _____
○ _____
○ _____

02 MONDAY

○ _____
○ _____
○ _____
○ _____
○ _____
○ _____
○ _____
○ _____

03 TUESDAY

○ _____
○ _____
○ _____
○ _____
○ _____
○ _____
○ _____
○ _____

04 WEDNESDAY

○ _____
○ _____
○ _____
○ _____
○ _____

March
2026

05 THURSDAY

06 FRIDAY

07 SATURDAY

08 SUNDAY

March
2026

09 MONDAY

○ _____
○ _____
○ _____
○ _____
○ _____
○ _____
○ _____
○ _____

10 TUESDAY

○ _____
○ _____
○ _____
○ _____
○ _____
○ _____
○ _____
○ _____

11 WEDNESDAY

○ _____
○ _____
○ _____
○ _____
○ _____
○ _____
○ _____
○ _____

12 THURSDAY

○ _____
○ _____
○ _____
○ _____
○ _____

March
2026

13 FRIDAY

14 SATURDAY

15 SUNDAY

16 MONDAY

17 TUESDAY

○ _____
○ _____
○ _____
○ _____
○ _____
○ _____
○ _____
○ _____

18 WEDNESDAY

○ _____
○ _____
○ _____
○ _____
○ _____
○ _____
○ _____
○ _____

19 THURSDAY

○ _____
○ _____
○ _____
○ _____
○ _____
○ _____
○ _____
○ _____

20 FRIDAY

○ _____
○ _____
○ _____
○ _____
○ _____

21 SATURDAY

22 SUNDAY

23 MONDAY

24 TUESDAY

March
2026

25 WEDNESDAY

○
○
○
○
○
○
○
○

26 THURSDAY

○
○
○
○
○
○
○
○

27 FRIDAY

○
○
○
○
○
○
○
○

28 SATURDAY

○
○
○
○
○

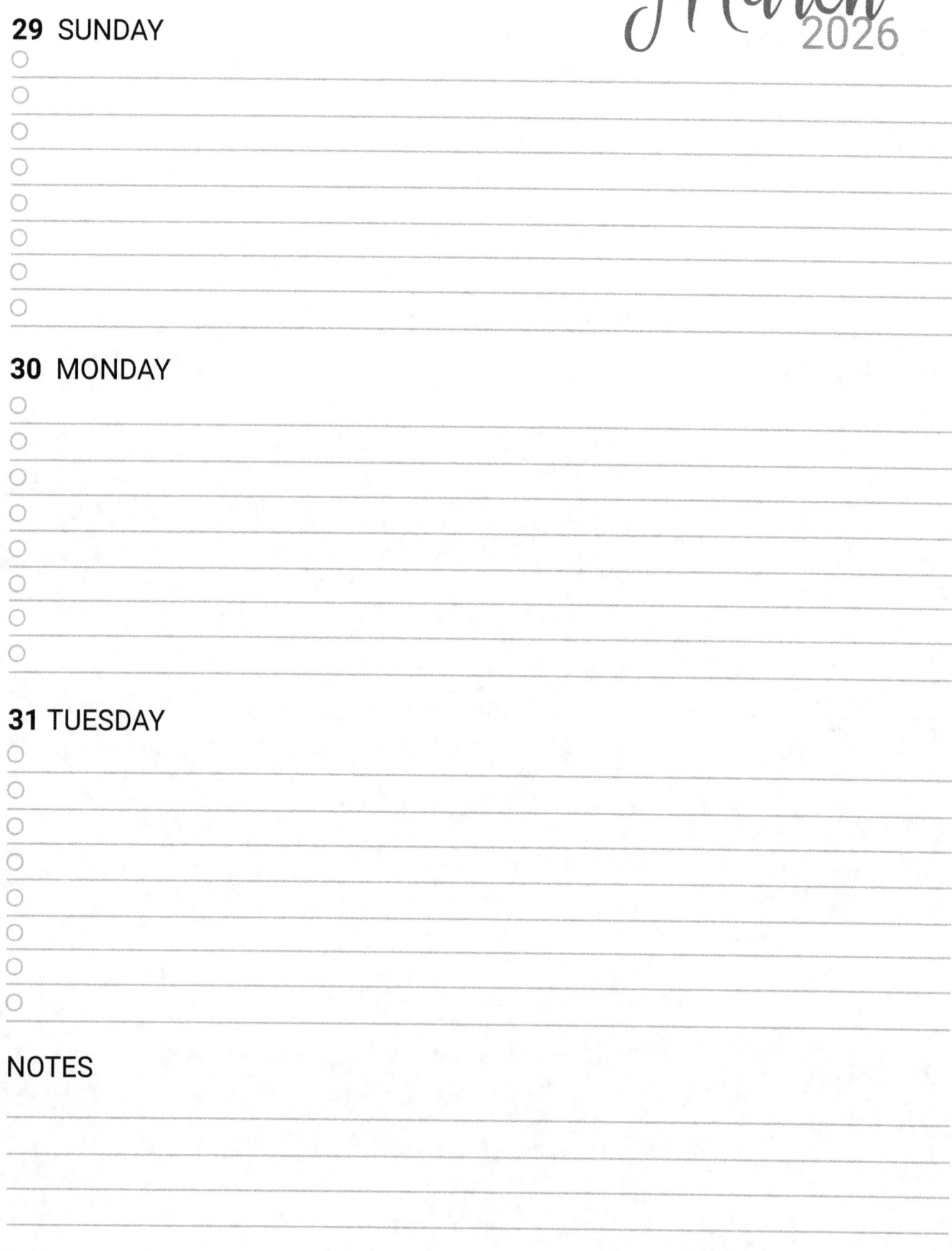

March
2026

29 SUNDAY

30 MONDAY

31 TUESDAY

NOTES

April
2026

01 WEDNESDAY

○
○
○
○
○
○
○
○

02 THURSDAY

○
○
○
○
○
○
○
○
○

03 FRIDAY

○
○
○
○
○
○
○
○

04 SATURDAY

○
○
○
○
○

April
2026

05 SUNDAY

06 MONDAY

07 TUESDAY

08 WEDNESDAY

April
2026

09 THURSDAY
○
○
○
○
○
○
○
○

10 FRIDAY
○
○
○
○
○
○
○
○

11 SATURDAY
○
○
○
○
○
○
○
○

12 SUNDAY
○
○
○
○
○

13 MONDAY

14 TUESDAY

15 WEDNESDAY

16 THURSDAY

17 FRIDAY

18 SATURDAY

19 SUNDAY

20 MONDAY

21 TUESDAY

22 WEDNESDAY

23 THURSDAY

24 FRIDAY

25 SATURDAY

26 SUNDAY

27 MONDAY

28 TUESDAY

29 WEDNESDAY

○ _____
○ _____
○ _____
○ _____
○ _____
○ _____
○ _____
○ _____

30 THURSDAY

○ _____
○ _____
○ _____
○ _____
○ _____
○ _____
○ _____
○ _____

NOTES

_____
_____
_____
_____
_____
_____
_____
_____
_____
_____
_____
_____

May
2026

01 FRIDAY
○ _____
○ _____
○ _____
○ _____
○ _____
○ _____
○ _____
○ _____

02 SATURDAY
○ _____
○ _____
○ _____
○ _____
○ _____
○ _____
○ _____
○ _____

03 SUNDAY
○ _____
○ _____
○ _____
○ _____
○ _____
○ _____
○ _____
○ _____

04 MONDAY
○ _____
○ _____
○ _____
○ _____
○ _____

May
2026

05 TUESDAY

06 WEDNESDAY

07 THURSDAY

08 FRIDAY

09 SATURDAY

10 SUNDAY

11 MONDAY

12 TUESDAY

13 WEDNESDAY

14 THURSDAY

15 FRIDAY

16 SATURDAY

17 SUNDAY
○
○
○
○
○
○
○
○

18 MONDAY
○
○
○
○
○
○
○
○
○

19 TUESDAY
○
○
○
○
○
○
○
○

20 WEDNESDAY
○
○
○
○
○

21 THURSDAY

22 FRIDAY

23 SATURDAY

24 SUNDAY

May
2026

25 MONDAY

26 TUESDAY

27 WEDNESDAY

28 THURSDAY

29 FRIDAY

30 SATURDAY

31 SUNDAY

NOTES

June 2026

01 MONDAY

- ○
- ○
- ○
- ○
- ○
- ○
- ○
- ○

02 TUESDAY

- ○
- ○
- ○
- ○
- ○
- ○
- ○
- ○

03 WEDNESDAY

- ○
- ○
- ○
- ○
- ○
- ○
- ○
- ○

04 THURSDAY

- ○
- ○
- ○
- ○
- ○

June 2026

05 FRIDAY

○
○
○
○
○
○
○

06 SATURDAY

○
○
○
○
○
○
○
○

07 SUNDAY

○
○
○
○
○
○
○
○

08 MONDAY

○
○
○
○
○

09 TUESDAY

○ _____
○ _____
○ _____
○ _____
○ _____
○ _____
○ _____
○ _____

10 WEDNESDAY

○ _____
○ _____
○ _____
○ _____
○ _____
○ _____
○ _____
○ _____

11 THURSDAY

○ _____
○ _____
○ _____
○ _____
○ _____
○ _____
○ _____
○ _____

12 FRIDAY

○ _____
○ _____
○ _____
○ _____
○ _____

13 SATURDAY

○
○
○
○
○
○
○
○

14 SUNDAY

○
○
○
○
○
○
○
○

15 MONDAY

○
○
○
○
○
○
○
○

16 TUESDAY

○
○
○
○
○

17 WEDNESDAY

18 THURSDAY

19 FRIDAY

20 SATURDAY

21 SUNDAY

○ _____
○ _____
○ _____
○ _____
○ _____
○ _____
○ _____
○ _____

22 MONDAY

○ _____
○ _____
○ _____
○ _____
○ _____
○ _____
○ _____
○ _____

23 TUESDAY

○ _____
○ _____
○ _____
○ _____
○ _____
○ _____
○ _____
○ _____

24 WEDNESDAY

○ _____
○ _____
○ _____
○ _____
○ _____

25 THURSDAY

○
○
○
○
○
○
○
○

26 FRIDAY

○
○
○
○
○
○
○
○

27 SATURDAY

○
○
○
○
○
○
○
○

28 SUNDAY

○
○
○
○
○

29 MONDAY

○ _____
○ _____
○ _____
○ _____
○ _____
○ _____
○ _____
○ _____

30 TUESDAY

○ _____
○ _____
○ _____
○ _____
○ _____
○ _____
○ _____
○ _____

NOTES

_____
_____
_____
_____
_____
_____
_____
_____
_____
_____
_____
_____
_____
_____

July
2026

01 WEDNESDAY

○
○
○
○
○
○
○
○

02 THURSDAY

○
○
○
○
○
○
○
○

03 FRIDAY

○
○
○
○
○
○
○
○

04 SATURDAY

○
○
○
○
○

July

05 SUNDAY

06 MONDAY

07 TUESDAY

08 WEDNESDAY

09 THURSDAY

10 FRIDAY

11 SATURDAY

12 SUNDAY

13 MONDAY

14 TUESDAY

15 WEDNESDAY

16 THURSDAY

17 FRIDAY

○
○
○
○
○
○
○
○

18 SATURDAY

○
○
○
○
○
○
○
○
○

19 SUNDAY

○
○
○
○
○
○
○
○

20 MONDAY

○
○
○
○
○

21 TUESDAY
○
○
○
○
○
○
○
○

22 WEDNESDAY
○
○
○
○
○
○
○
○

23 THURSDAY
○
○
○
○
○
○
○
○

24 FRIDAY
○
○
○
○
○

25 SATURDAY

○ _____
○ _____
○ _____
○ _____
○ _____
○ _____
○ _____
○ _____

26 SUNDAY

○ _____
○ _____
○ _____
○ _____
○ _____
○ _____
○ _____
○ _____

27 MONDAY

○ _____
○ _____
○ _____
○ _____
○ _____
○ _____
○ _____
○ _____

28 TUESDAY

○ _____
○ _____
○ _____
○ _____
○ _____

July
2026

29 WEDNESDAY

30 THURSDAY

31 FRIDAY

NOTES

August
2026

01 SATURDAY

○ _____
○ _____
○ _____
○ _____
○ _____
○ _____
○ _____
○ _____

02 SUNDAY

○ _____
○ _____
○ _____
○ _____
○ _____
○ _____
○ _____
○ _____

03 MONDAY

○ _____
○ _____
○ _____
○ _____
○ _____
○ _____
○ _____
○ _____

04 TUESDAY

○ _____
○ _____
○ _____
○ _____
○ _____

05 WEDNESDAY

06 THURSDAY

07 FRIDAY

08 SATURDAY

August
2026

09 SUNDAY

○
○
○
○
○
○
○
○

10 MONDAY

○
○
○
○
○
○
○
○

11 TUESDAY

○
○
○
○
○
○
○
○

12 WEDNESDAY

○
○
○
○
○

13 THURSDAY

14 FRIDAY

15 SATURDAY

16 SUNDAY

17 MONDAY

18 TUESDAY

19 WEDNESDAY

20 THURSDAY

August
2026

21 FRIDAY

22 SATURDAY

23 SUNDAY

24 MONDAY

25 TUESDAY

26 WEDNESDAY

27 THURSDAY

28 FRIDAY

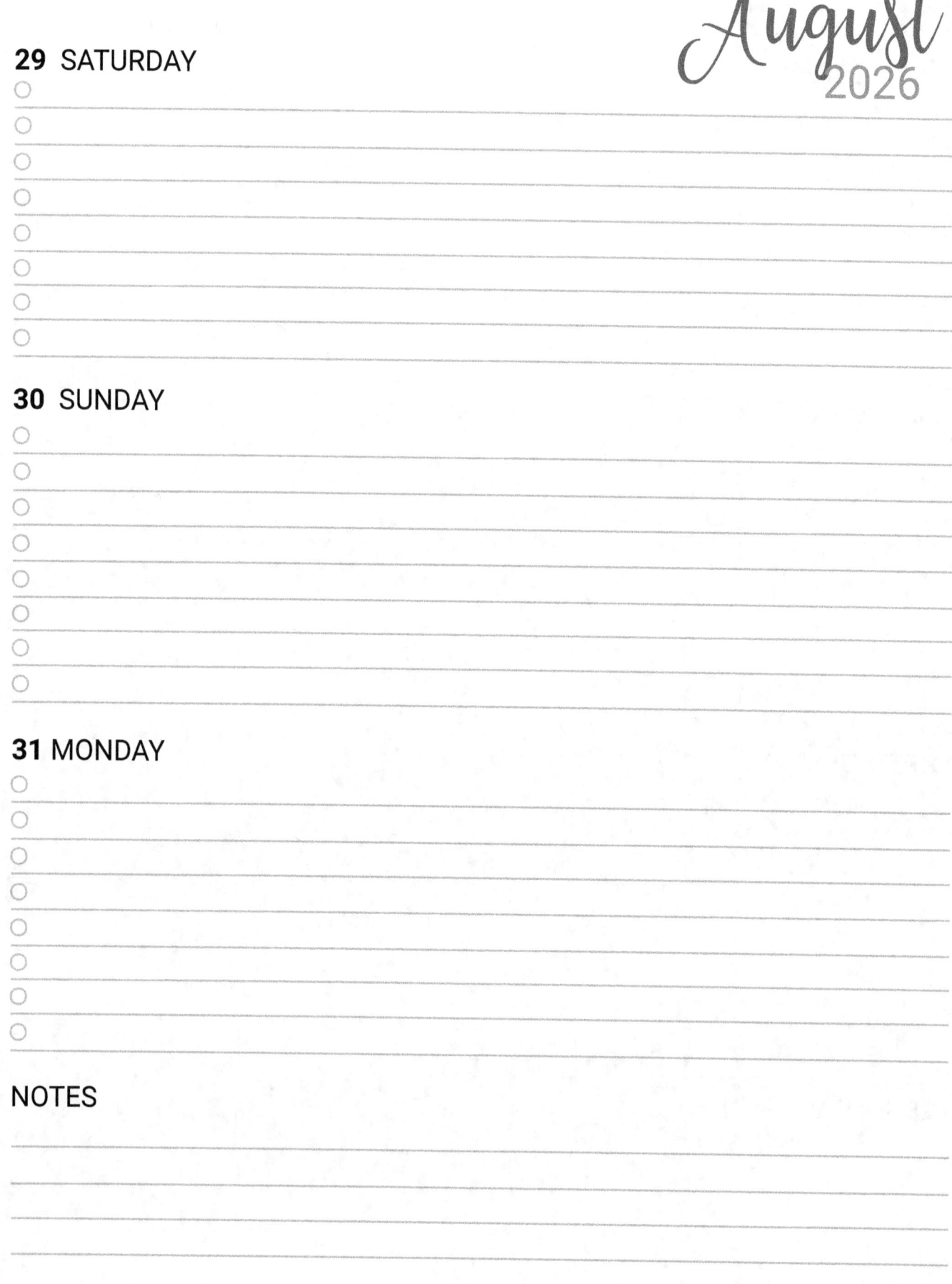

August
2026

29 SATURDAY

30 SUNDAY

31 MONDAY

NOTES

01 TUESDAY

○ _____
○ _____
○ _____
○ _____
○ _____
○ _____
○ _____
○ _____

02 WEDNESDAY

○ _____
○ _____
○ _____
○ _____
○ _____
○ _____
○ _____
○ _____

03 THURSDAY

○ _____
○ _____
○ _____
○ _____
○ _____
○ _____
○ _____
○ _____

04 FRIDAY

○ _____
○ _____
○ _____
○ _____
○ _____

05 SATURDAY

○ _____
○ _____
○ _____
○ _____
○ _____
○ _____
○ _____
○ _____

06 SUNDAY

○ _____
○ _____
○ _____
○ _____
○ _____
○ _____
○ _____
○ _____

07 MONDAY

○ _____
○ _____
○ _____
○ _____
○ _____
○ _____
○ _____
○ _____

08 TUESDAY

○ _____
○ _____
○ _____
○ _____
○ _____

09 WEDNESDAY

○
○
○
○
○
○
○
○

10 THURSDAY

○
○
○
○
○
○
○
○

11 FRIDAY

○
○
○
○
○
○
○
○

12 SATURDAY

○
○
○
○
○

13 SUNDAY

14 MONDAY

15 TUESDAY

16 WEDNESDAY

17 THURSDAY

18 FRIDAY

19 SATURDAY

20 SUNDAY

21 MONDAY

○
○
○
○
○
○
○
○

22 TUESDAY

○
○
○
○
○
○
○
○

23 WEDNESDAY

○
○
○
○
○
○
○
○

24 THURSDAY

○
○
○
○
○

25 FRIDAY

- ○
- ○
- ○
- ○
- ○
- ○
- ○
- ○

26 SATURDAY

- ○
- ○
- ○
- ○
- ○
- ○
- ○
- ○

27 SUNDAY

- ○
- ○
- ○
- ○
- ○
- ○
- ○
- ○

28 MONDAY

- ○
- ○
- ○
- ○
- ○

29 TUESDAY

○ _____
○ _____
○ _____
○ _____
○ _____
○ _____
○ _____
○ _____

30 WEDNESDAY

○ _____
○ _____
○ _____
○ _____
○ _____
○ _____
○ _____
○ _____

NOTES

_____
_____
_____
_____
_____
_____
_____
_____
_____
_____
_____
_____

October
2026

01 THURSDAY

○
○
○
○
○
○
○
○

02 FRIDAY

○
○
○
○
○
○
○
○
○

03 SATURDAY

○
○
○
○
○
○
○
○

04 SUNDAY

○
○
○
○
○

October
2026

05 MONDAY

○
○
○
○
○
○
○
○

06 TUESDAY

○
○
○
○
○
○
○
○

07 WEDNESDAY

○
○
○
○
○
○
○
○

08 THURSDAY

○
○
○
○
○

October
2026

09 FRIDAY
○
○
○
○
○
○
○
○

10 SATURDAY
○
○
○
○
○
○
○
○

11 SUNDAY
○
○
○
○
○
○
○
○

12 MONDAY
○
○
○
○
○

13 TUESDAY

○
○
○
○
○
○
○
○

14 WEDNESDAY

○
○
○
○
○
○
○
○

15 THURSDAY

○
○
○
○
○
○
○
○

16 FRIDAY

○
○
○
○
○

October
2026

17 SATURDAY

○

○

○

○

○

○

○

○

18 SUNDAY

○

○

○

○

○

○

○

○

19 MONDAY

○

○

○

○

○

○

○

○

20 TUESDAY

○

○

○

○

○

21 WEDNESDAY

22 THURSDAY

23 FRIDAY

24 SATURDAY

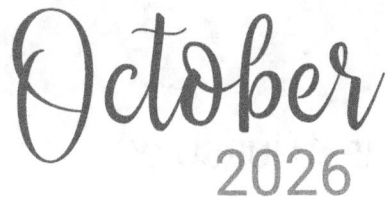

October
2026

25 SUNDAY

○
○
○
○
○
○
○
○

26 MONDAY

○
○
○
○
○
○
○
○

27 TUESDAY

○
○
○
○
○
○
○
○

28 WEDNESDAY

○
○
○
○
○

29 THURSDAY

30 FRIDAY

31 SATURDAY

NOTES

November
2026

01 SUNDAY

○ _____
○ _____
○ _____
○ _____
○ _____
○ _____
○ _____
○ _____

02 MONDAY

○ _____
○ _____
○ _____
○ _____
○ _____
○ _____
○ _____
○ _____

03 TUESDAY

○ _____
○ _____
○ _____
○ _____
○ _____
○ _____
○ _____
○ _____

04 WEDNESDAY

○ _____
○ _____
○ _____
○ _____
○ _____

November
2026

05 THURSDAY

○ _____
○ _____
○ _____
○ _____
○ _____
○ _____
○ _____
○ _____

06 FRIDAY

○ _____
○ _____
○ _____
○ _____
○ _____
○ _____
○ _____
○ _____

07 SATURDAY

○ _____
○ _____
○ _____
○ _____
○ _____
○ _____
○ _____
○ _____

08 SUNDAY

○ _____
○ _____
○ _____
○ _____
○ _____

09 MONDAY

10 TUESDAY

11 WEDNESDAY

12 THURSDAY

November
2026

13 FRIDAY

14 SATURDAY

15 SUNDAY

16 MONDAY

November
2026

17 TUESDAY
○ _____
○ _____
○ _____
○ _____
○ _____
○ _____
○ _____
○ _____

18 WEDNESDAY
○ _____
○ _____
○ _____
○ _____
○ _____
○ _____
○ _____
○ _____

19 THURSDAY
○ _____
○ _____
○ _____
○ _____
○ _____
○ _____
○ _____
○ _____

20 FRIDAY
○ _____
○ _____
○ _____
○ _____
○ _____

November
2026

21 SATURDAY

22 SUNDAY

23 MONDAY

24 TUESDAY

November
2026

25 WEDNESDAY

○
○
○
○
○
○
○
○

26 THURSDAY

○
○
○
○
○
○
○
○

27 FRIDAY

○
○
○
○
○
○
○
○

28 SATURDAY

○
○
○
○
○

29 SUNDAY

○
○
○
○
○
○
○
○

30 MONDAY

○
○
○
○
○
○
○
○

NOTES

December
2026

01 TUESDAY

○
○
○
○
○
○
○
○

02 WEDNESDAY

○
○
○
○
○
○
○
○

03 THURSDAY

○
○
○
○
○
○
○
○

04 FRIDAY

○
○
○
○
○

December
2026

05 SATURDAY

06 SUNDAY

07 MONDAY

08 TUESDAY

09 WEDNESDAY

10 THURSDAY

11 FRIDAY

12 SATURDAY

13 SUNDAY

14 MONDAY

15 TUESDAY

16 WEDNESDAY

17 THURSDAY

18 FRIDAY

19 SATURDAY

20 SUNDAY

21 MONDAY

22 TUESDAY

23 WEDNESDAY

24 THURSDAY

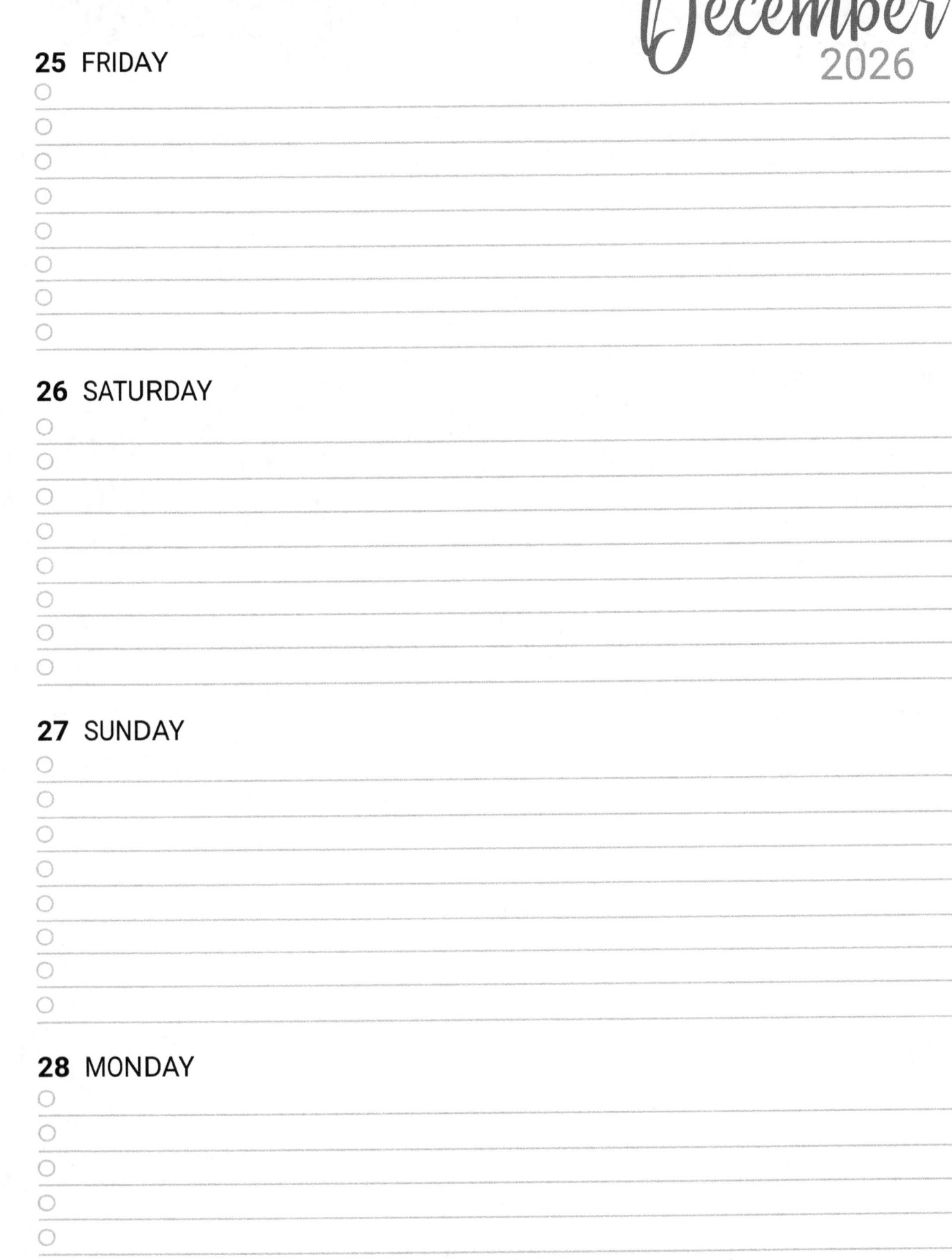

December
2026

25 FRIDAY

26 SATURDAY

27 SUNDAY

28 MONDAY

December
2026

29 TUESDAY

○
○
○
○
○
○
○
○

30 WEDNESDAY

○
○
○
○
○
○
○
○

31 THURSDAY

○
○
○
○
○
○
○
○

NOTES

25 FRIDAY

○
○
○
○
○
○
○
○

26 SATURDAY

○
○
○
○
○
○
○
○

27 SUNDAY

○
○
○
○
○
○
○
○

28 MONDAY

○
○
○
○
○

29 TUESDAY

○
○
○
○
○
○
○
○

30 WEDNESDAY

○
○
○
○
○
○
○
○

31 THURSDAY

○
○
○
○
○
○
○
○

NOTES